science
on display

Carolyn Dale

Acknowledgements

The author and publisher would like to thank the children and staff of the following schools for their hard work, co-operation and expertise in producing the displays for this book:

All Saints Junior School, Maidenhead
Boyn Hill Infant School, Maidenhead
Lent Rise Primary School, Burnham
Lowbrook Primary School, Cox Green, Maidenhead

The author would particularly like to thank Janet Adkins, Judith Castle, Shirley Craddock and Sylvia Harris for their enthusiasm, expertise and imaginative ideas in helping children to produce the work for this book. Finally, thanks go to Zoë Nichols and Kelvin Freeman for the photo-shoot days in which their expertise was invaluable.

Oats and Beans and Barley Grow (page 16)

First published in 2005 by Belair Publications.
Apex Business Centre, Boscombe Road, Dunstable, LU5 4RL.

Commissioning Editor: Zoë Nichols   Editor: Jennifer Steele
Page Layout: Suzanne Ward   Photography: Kelvin Freeman
Cover Design: Steve West

© 2005 Folens on behalf of the author.

British Library Cataloguing in Publication Data. A catalogue record for this publication is available from the British Library.

ISBN 0 94788 277 4

Contents

Introduction

Welcome to *Science on Display*. Perhaps the two are not normally spoken of in the same breath, but display is an essential part of developing children's understanding in Science, helping them to picture their ideas, to learn about the ideas of others and to motivate them.

Teaching and Learning Science

Science can be difficult to define but it is about trying to understand the things around us:

- the environment and the living things in it; the vast variety; how plants and animals live and adapt to life in their habitats and how they interact together

- the different things that can benefit or destroy environments

- materials, both natural and manufactured; where they are found and how they are made; how their properties suit them to their uses

- the physical world of electricity, forces, energy and the Universe.

Science is about our everyday lives. Some scientific ideas are difficult to comprehend, but understanding does depend to a great degree on how we are taught. It is said that Science is not so much about finding answers, but more about asking questions. It could be argued that there are some golden rules about teaching and learning in Science:

- **enthuse the children** by using contexts that they enjoy, such as music, sports, food, clothes, holidays, homes, animals and so on. Use their own environment – the city centre, the forest, the sea, wherever they live. Get them out, let them observe at first hand and ask questions about it

- **find out the children's ideas** by using techniques that allow them to show or tell us what they think now. This may be by using a concept cartoon, asking them to construct a concept map, or getting them to tell or show through words, drawings or drama

- **use their ideas to inform teaching by** challenging children and allowing them to test their ideas. This means using their experiences to make predictions first before testing

- **encourage the children to ask questions** about things and find ways of answering them. Help them to understand how to formulate investigative questions, using 'how', 'what will happen', 'which'. 'Why' questions tend to be those that require the use of secondary sources. Try to keep questions open rather than closed – open questions allow for a range of answers rather than an expected 'right' answer

- **discuss and evaluate results and methods**, and encourage the children to think how they might improve what they did to make the results more valid.

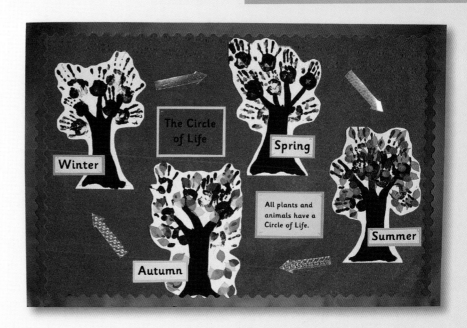

How Do Displays Help in Science?

Children love to create, and to use language, materials and techniques to represent their ideas and experiences. Sometimes they can't express their ideas as well as they would like to in simple language. Displays can therefore help children to:

- think, suggest ideas and present findings imaginatively
- express themselves in ways that are appropriate to them
- share ideas and work co-operatively; listen to the ideas of others.

Science displays usually include other areas of the curriculum. There are clear links with Art and Design, and Design and Technology. But literacy skills are also often required, as are the mathematical skills of measuring, knowledge of shapes and data handling. And we should not overlook the all-important development of positive attitudes of working together towards an end product.

What Is in This Book?

Thirty-three science themes appropriate for the whole Primary age range are covered in this book. Each theme contains:

- **A Clear Focus of Learning** which highlights the focus for the display and the following activities.

- **Starting Points** which introduce the main scientific concept through whole class discussion and practical activities.

- **Displays** that have been designed to introduce the children to the Science concept and record their ideas about it.

- **Further Activities** are practical activities that carry on the theme, with suggestions of how to introduce questions to promote good scientific activity and thinking.

This book aims to give a range of ideas to meet the needs of the curriculum, but ideas contained within each theme can also be adapted to suit your learning objectives and the various stages and abilities of the children.

Moving Hands

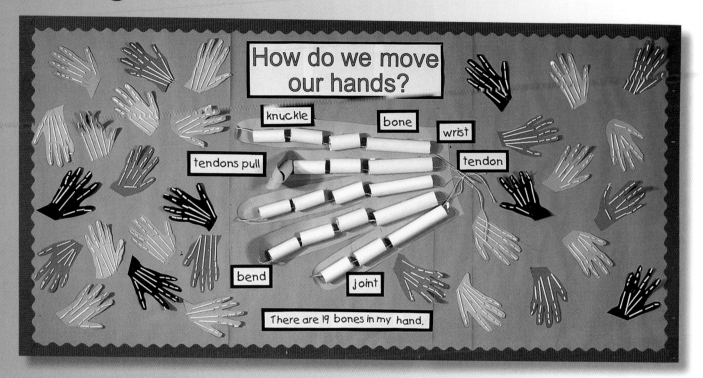

How do we move our hands?

knuckle   bone   wrist

tendons pull   tendon

bend   joint

There are 19 bones in my hand.

Starting Points

- Look at a bone from any animal. Ask the children to describe it. Why do animals have skeletons? Introduce the human skeleton. Look at pictures of a model skeleton and at X-rays. Discuss the purposes of a skeleton – for movement, protection and shape and support of the body. Discuss that bones are living tissue.

- Identify places where we bend. Talk about joints and muscles and how they help us move.

- Look closely at our own hands. How many bones are inside? What are the clues? Do bigger hands have more bones in them?

Focus of Learning

Examining the structure of a human skeleton
Understanding how bones, muscles and joints help us to move

Making the Display

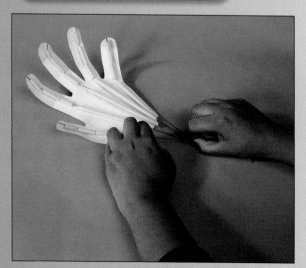

- Create a large hand for the centre, using cardboard tubes for each section.

- Attach strings to the fingertips and thread through the tubes to the wrist. The strings will be pulled, so the ends need to be firmly attached to the ends of each finger.

- Ask the children to draw round their own hands on to thin card and cut out. Choose colours to represent different skin colours.

- Bend the hand shape inwards to match the bends in a hand.

- Cut out pieces of straw to match the length of each section between the joints. Stick down and attach the strings as for the main hand.

- Pull the strings, so that the fingers and thumb move. Discuss what each part represents. What sorts of movements can the children do with the model hand? Compare this with their own hand. Which is better and how?

Further Activities

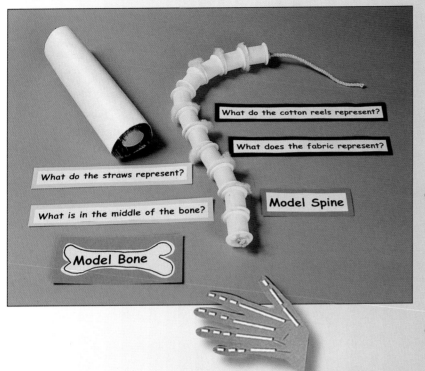

What do the cotton reels represent?

What does the fabric represent?

What do the straws represent?

What is in the middle of the bone?

Model Spine

Model Bone

- Make a model of the inside of a bone. Use a cylinder of thin card or paper and a smaller one to fit inside. Fill the gap with straws. Discuss with the children what each part represents (the outer card represents the compacted outer bone; the straws represent the spongy inside of a bone through which nerves and blood vessels pass; the hollow centre represents the bone marrow cavity, which would contain jelly-like bone marrow that produces red blood cells). Why are bones not solid? What is in the centre gap?

- Where do we bend? On a large sheet of paper, draw around the outline of a child lying on the floor. Bend the outline at each of the major joints and name them. Ask the children to move as if they did not have joints. Draw the main bones inside the outline to show where the bones join.

- Create movement pictures using collage materials. Add a 3D effect using Modroc plaster bandages.

- Make a model arm. Cut out the shape and size of the bones in the upper and lower arm from thick card. Attach elastic bands to represent the biceps and triceps muscles. Pull the arm down and note that the triceps stretch. Pull the arm up and the biceps pull. Use this to explain that muscles can only pull, they cannot push.

- Make a model skeleton using art straws and paper to include one section for the skull, one for the ribs and one for the hip girdle. Make the arms and legs from straws. Discuss how a skeleton protects vital parts of our bodies.

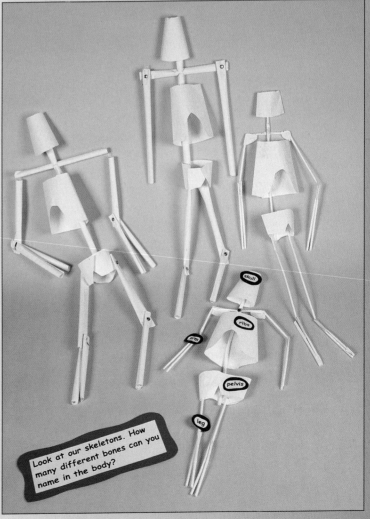

skull

ribs

arm

pelvis

leg

Look at our skeletons. How many different bones can you name in the body?

Drain Creatures

COULD THESE CREATURES SURVIVE IN OUR DRAINS ?

dingy · musty · unhygienic · smelly · damp · dark · cold · clubbybug · Galasad · clubbybug (young) · Hoddy · foul · Gulco · squalid · mouldy · Elephanfish · Sewer ray · Flubber · TeTree

Starting Points

- Discuss the term 'habitat'. List the habitats of given animals and plants and where they live. How are the life processes of a particular animal suited to its habitat?

- Introduce the habitat of a drain. Brainstorm the conditions there; use vocabulary such as organism, algae, microbes/micro-organisms, habitat. Design organisms that could live in the drain.

Making the Display

- Create a drain habitat on a large board; use a dark background and transparent plastic to show water going into the drain and at the bottom.

- Create creatures from junk materials based on the children's ideas. Attach to the board, along with pictures of debris.

- Add the names of creatures, which could suggest something about their life processes.

Focus of Learning

Exploring how animals are suited to their habitats

Understanding the components of food chains

Further Activities

- Ask the children to make a fact sheet about their creature, explaining how its life processes suit its drain habitat.

- Sort some animals using a Carroll Diagram. Devise questions from the diagram. Why are there no animals in one of the sections? Which animals have wings but do not fly? Alternative headings could be: legs/no legs; lays eggs/does not lay eggs; six legs/not six legs; feeds on mother's milk/does not feed on mother's milk.

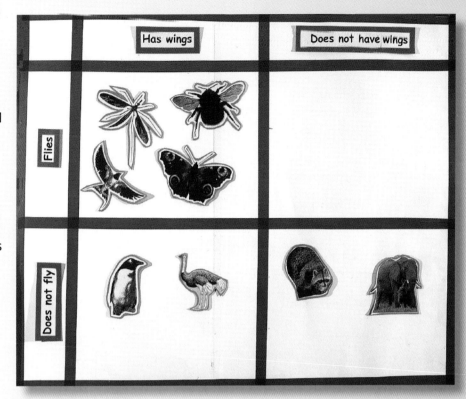

	Has wings	Does not have wings
Flies		
Does not fly		

How have our butterflies adapted to their environment?
Why have they done this?

- Look at the parts of different animals used in movement, for example the wing of a bird/bat; the flipper of a whale/foot of a frog. How is each adapted to its habitat?

- Talk about camouflage. Make rubbings of tree barks. Draw two outlines of a butterfly/moth. Colour one so that it will not be found on the bark and the other so that it will show up well in the same place. Find out about peppered moths and their colouration, and where they survive.

- Look at food chains. Construct hanging mobiles of food chains with at least three components and use vocabulary such as producer and consumer. Discuss what will happen to a food chain if the producer or consumer dies out.

FOOD CHAIN

leaf

bird

Ron Rotten Tooth

Focus of Learning

Learning about teeth and how to care for them

Starting Points

- Ask the children to show you their best smiles. Do this at the beginning of the day and after lunch. Tell them that after lunch their smiles are not as good as they were in the morning. Why?

- Introduce the character of Ron Rotten who has bad teeth and a really ugly smile. Make up a script for Ron, who tells what he eats and how he treats his teeth.

- Discuss decay. Can we brush it off our teeth? How can we ensure that our teeth stay healthy? Talk about visiting the dentist regularly.

Making the Display

- Prepare the background using flesh-coloured paper.

- Stick on a large open mouth. Make 3D eyes, nose and hair, and stick on the face. Make the expression woeful!

- Create model teeth from small boxes, shaped to represent different teeth. Link to work on nets in Maths. The children should recognise the different shapes of teeth and copy them as accurately as possible.

- Decorate each tooth as rotten or healthy. Most will be rotten. Stick the teeth into the mouth.

- Add a large toothbrush and toothpaste cut from card, and a sign about looking after teeth.

Further Activities

- Talk about how snacks affect teeth. Make two stand-up teeth from thin white card. Choose a variety of foods, some of which are 'OK' for our teeth and others that can cause decay ('DK'). Stick pictures of food on to the correct tooth. Stress to the children that they should brush their teeth after eating fruit as a snack because fruits contain a lot of sugar, although eating fruit has lots of health benefits.

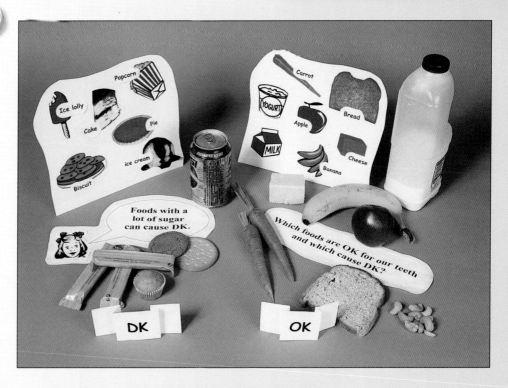

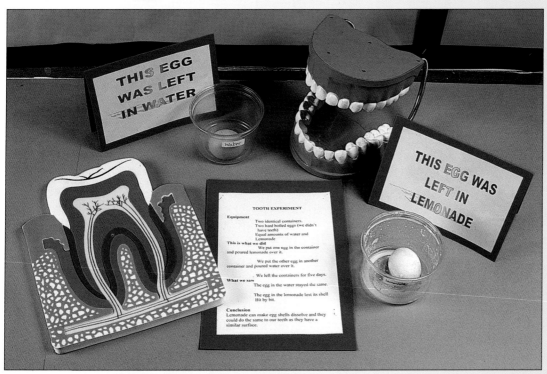

- What can the children find out about teeth? Collage a diagram of a tooth to show the structure. Show a diagram or model of an adult mouth with permanent teeth; include top and bottom sets for children to label and find out the purpose of each type of tooth.

- Discuss the effects of fizzy drinks on teeth. Boil two eggs. The eggs in their shells represent teeth, the shell being the enamel. Put one in water and the other in a fizzy drink. Leave and observe. Keep a record of changes over a period of weeks. Explain that enamel is the hardest substance in the human body – and discuss what fizzy drinks do to it!

- Devise a play entitled 'Ron's Bad Dream'. Show what Ron eats before going to bed and that he does not brush his teeth. Show his dream in which bacteria attack his teeth while he is asleep. Make the bacteria really ugly and vicious! Introduce Nancy Newtooth as a new, bright and shiny tooth. Show how Nancy may be kept bright and shiny, for example by remembering to clean teeth and visiting the dentist regularly.

What Can Micro-organisms Do?

- Children may have strange ideas about some kinds of micro-organisms, even thinking that they are some kind of alien being. Ask groups to record their ideas – where they are found, different types and what they do. Keep their recordings for future reference.

- Talk about the ways in which micro-organisms can be dangerous but also introduce the idea of ways in which they can help us.

- Micro-organisms may also be called microbes, but 'germ' is not an accurate alternative. Yeast is a useful microbe. Introduce others that are helpful and some that are not.

Focus of Learning

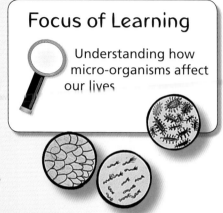

Understanding how micro-organisms affect our lives

Making the Display

⚠ **Rotting food must always be kept in sealed bags or other sealed containers and not opened, then disposed of in the same containers.**

- Create a background, split into two – one side for helpful microbes and the other for unhelpful microbes.

- Accurately draw some enlarged micro-organisms on discs of paper. In the centre of the display board attach the title as a question: What can micro-organisms do?

- Organise the children into five groups to represent one area of the following: microbes make us ill – paint or collage a person with spots and holding their stomach and groaning; microbes make food mouldy – include samples of mouldy bread sealed in transparent bags; microbes make food – attach pictures of different foods, such as bread, beer, wine, pickles, yoghurt, cheese; microbes make compost – attach a picture of a pile of compost in a bin; microbes clean sewage – add a picture of a sewage farm/tank with bacteria at the bottom of sludge.

- Attach all the elements to the correct sides of the display board.

Further Activities

- Compare the rates at which white and brown breads (and other types) become mouldy.

- Make a water microscope using a glass and a drop of water. Try other liquids to find out which is the most effective. Compare the water microscope with a real one.

- Choose at least five different types of food. Put them in sealed containers. Predict how they will change over two weeks. Test to find out. Why do the children think some foods decayed more than others?

- Talk about the ways in which disease can be prevented. Make leaflets about scientists such as Louis Pasteur, Edward Jenner and John Snow, who have helped in the fight against disease.

- Make a compost tower using four lemonade bottles. Bottle 1: cut a few centimetres below its widest point. Punch air holes in the cap. Bottle 2: cut out a long middle section and fit into bottle 1. Cut four air vents in the sides and cover with a mesh. Bottle 3: cut just above the widest point at the bottom. Invert and fit into bottle 2. Punch holes in the lid and put vents in the sides as in bottle 2. Bottle 4: cut at the widest point at the top and fit into bottle 3. The holes covered in mesh are most important to let air in. Use old tights for the mesh and for covering the punched lids. Fill the tower with plant pieces and watch to see what happens over time.

A Walk on the Wild Side

Starting Points

- Conduct a survey outside and list the minibeasts found locally.

- Collect some minibeasts humanely and observe in detail in the classroom. Ensure that the animals are kept in their preferred conditions and with sufficient food. Use hand lenses to observe the structure of the minibeasts and their habits. Find out about their habitats and the conditions there – light/dark, damp/dry, warm/cold and so on.

- Talk about how the conditions affect the animals living there.

Focus of Learning

Learning about how minibeasts have adapted to different habitats
Recognising that a key is a device for sorting animals

Making the Display

- Create a winding track across the board. Decorate around the track with different types of habitat, for example pond, river, forest, playing fields, paving, gardens, sandy area, bare rock, cabbage field, privet hedge. Make the habitats 3D to represent the various conditions, for example use shredded paper to represent grassy areas and so on.

- Label some of the squares near each habitat with the relevant conditions.

- Make animals for each habitat, for example woodlouse, ant, ladybird, bee, worm, snail. Place these on the display board in the correct habitats.

Further Activities

- Look at a passport and list the type and form of information on it. Design one for an animal, giving the necessary information about features and habitat. Research the information on the Internet or in books.

- Create addresses for minibeasts. Write them clearly on an envelope and add an imaginative letter or invitation to put inside. Also include information about the sender.

- Make accurate models of one of the animals using a variety of materials. It is essential if it is an insect, that it has three body parts, six legs attached to the thorax, the correct number of wings also attached to the thorax and so on. Find other animals that are similar to each model by using secondary sources.

- Select a particular habitat worldwide, for example desert or rainforest, and find out about the conditions in each. Research the mini-habitats within each larger one and compare the animals living there. Compare, for example, the animals and conditions at ground level in a rainforest with those at tree level.

- Devise a key to sort out six minibeasts. It is essential that all questions be 'closed' with an answer of 'yes' or 'no'. Attach accurate drawings and names of animals and use scientific language where possible. Test each other's keys to check if they work. Devise other keys using animals from a particular habitat, for example a pond.

Our key

Does it have legs?

Does it have wings?   Does it have a shell?

Does it have a sting?   snail   Does it have segments?

spider

butterfly

bee   worm   slug

Oats and Beans and Barley Grow

Focus of Learning

Learning what grains and seeds need to grow into healthy plants

Starting Points

- Examine grains and beans with the children. Describe them in detail. What is inside? Observe them with hand lenses.

- Find out the children's ideas of what the grains need to make them germinate by asking them each to draw a picture. If they think the grains need light then ask them to plan how they might test this prediction.

- Find out about how a farmer grows wheat and barley and cares for the crop.

Making the Display

- Make a background of a field with 3D ears of corn growing. These should be as accurate as possible, based on observation of real grains. Put in the Sun, rain and clouds.

- Make a large 3D scarecrow using any suitable materials, for example a variety of grains. Add real wheat or other crops if available.

- Make crows with flapping wings. Stick a length of straw in the centre of the body. Attach strings to the wings and thread through the straw. Pull the strings to flap the wings. Attach to the display.

- Add the title, labels and questions concerning what grain needs to grow.

Further Activities

- Place grains of wheat/barley in trays and give them different amounts of water every other day. Keep a diary of growth rates. How many grains did the children put in each tray? How many of them germinated? Grow similar trays of grains in different amounts of light and monitor the growth rate. Also compare different types of grain growth. Do some grow better than others? Do some grow quicker, taller or have more leaves?

How much water do wheat grains need to grow well?

Can you make a noisy bird scarer?

- Tell 'The Wheat Grains' story. Ask the children to act out the parts of the scarecrow, crows, Sun, rain and wheat seeds to show how the crop grows. Include how to look after the wheat to make sure it grows into a healthy crop.

- Make some flour. Weigh an amount of wheat or barley grains. Grind the grains between stones/bricks to make flour. How much flour has been made? How is it different from the flour bought in shops?

- Scare the crows! Consider how the crop may be protected from crows, such as by sound, sight and smell. Design and make a bird scarer. How could its effectiveness be tested?

The Dandelion Clock

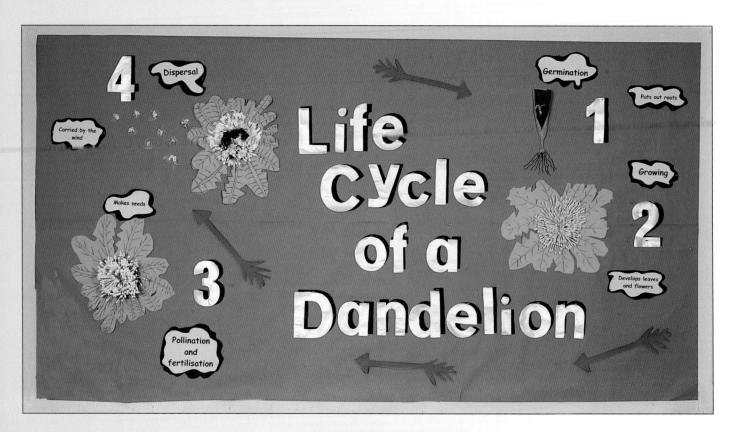

Starting Points

- Collect some complete dandelions, including the roots. Ask the children to make accurate labelled drawings of them, explaining the function of each of the main parts. Look carefully at the shape, arrangement and structure of the leaves and florets.

- Compare a dandelion clock and a flower head. Discuss the changes and fit them into a life cycle. Name the main stages of the life cycle – germination, growth, pollination, fertilisation, seed formation, dispersal.

- Ask groups of children to make up life-cycle cards for different familiar plants, such as buttercup, sunflower, poppy and so on. Challenge other groups to put the life-cycle cards in order. Compare and contrast the different stages.

Focus of Learning

Studying the life cycle of familiar plants and animals
Recognising that different conditions affect plant growth

Making the Display

- Provide a plain background. Look carefully at a single dandelion seed. Using a single cotton bud, represent one of the seeds germinating.

- Make a complete flower head using cotton buds painted yellow. Attach it to a background of cardboard leaves.

- Make a head using unpainted cotton buds showing the seeds.

- Make another head showing the seeds leaving the head.

- Link the stages with red arrows.

- Label each stage, using scientific terms and state what happens at each stage.

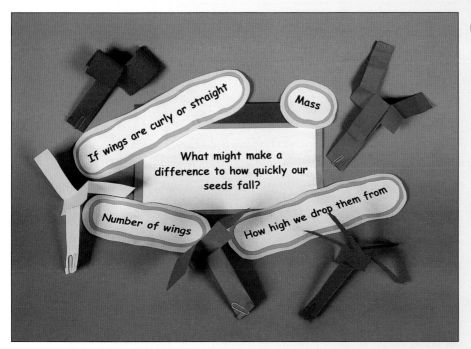

- Compare the height of dandelions in different habitats, for example in a field, hedgerow and under trees. List the different conditions in each, especially the amount of light each plant receives. Why do dandelions grow tall when there is little light?

- Compare the height of dandelions in an area that is regularly cut, one that is cut infrequently and an area where the grass is never cut.

- Dandelion seeds float like parachutes. Look at a parachute. Use a button to represent a seed and attach a paper parachute so that the 'seed' will float down slowly. Discuss why it is useful for seeds to have a parachute. How does the size of the parachute affect how quickly it falls?

- Look at other seeds that are light and float in the air, such as Old Man's Beard and Rosebay Willow Herb. Compare the structure of them.

- Look at other seeds that fall slowly, moving away from the parent plant by spinning, such as Sycamore and Ash keys. Make some spinning seeds from paper to the pattern of an autogyro. Drop the autogyro-seed and observe how it falls. Make up different designs for a spinning seed, using a variety of materials.

- Create a hanging mobile to show the life cycles of a frog, butterfly, dragonfly and other animals. Use wire coat hangers to make a circular-shaped frame with a single life stage at the end of each coat hanger. Compare the life stages of these animals with those of the plants studied.

Plants We Eat

Starting Points

- Make a collection of different foods. Which ones grow in the ground and are plants? The children could draw where they think they grow – underground/on trees or on bushes and so on.

- Sort out the fruits from a collection of fruit and vegetables. Ask groups of children to make a list of observations. Look inside each fruit. Encourage detailed observation of the number and position of seeds. Provide hand lenses. The children need to realise that all the plant parts that have seeds/stones inside are classified as fruits – so some fruits are not always sweet.

- Conduct two class surveys to find out the favourite and least favourite plants that the class eats. Show the data on a chart. Talk about the importance of eating fruit and vegetables.

Making the Display

- Make prints of fruit and vegetables. Select some of each to put together into sets.

- Create an awning for a fruit and vegetable stall, with paper rolls for uprights and colourful folded paper for the awning.

- Make the awning stand out from the background by sticking stiff cardboard triangles underneath. Put a title on the awning.

- Attach plant prints to the background. If linking to Maths, add some price labels.

- Put real samples on the table in front of the prints.

Focus of Learning

Understanding that eating different plants helps to keep us healthy Recognising what plants need to grow strong and healthy

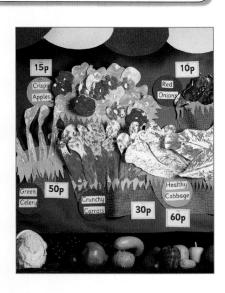

Further Activities

- To demonstrate what a plant needs to grow, cut out strips of three eggbox sections and put damp tissues or some soil into each and add some cress seeds. Decorate the front end as a caterpillar head. Discuss with the children how they might make the seeds grow. Water the cress, place in the light and watch a hairy caterpillar appear. Count the number of seeds planted and how many grow. Use the cress to make a sandwich.

What do seeds need to start to grow?

water ✓
light ✓
warmth ✗

water ✗
light ✓
warmth ✓

water ✓
light ✓
warmth ✓

water ✓
light ✗
warmth ✓

- Read *Avocado Baby* by John Burningham (published by Red Fox). Look at a real avocado. Cut it open and discuss that the stone is a seed. Try to grow it, giving it all the things it needs.

- Give each child an outline of the trunk of an apple tree. Complete the tree with handprints to show what it is like in each of the seasons – bare branches, with leaves, with blossom and with fruit. Arrange the pictures to show a cycle.

- Grow different vegetable seeds in biodegradable pots. Wind newspaper around a kitchen roll tube. Seal the edges with masking tape. Place a criss-cross of masking tape over the bottom of the newspaper pot. Remove the kitchen roll tube. Fill the newspaper pot with compost. Plant the seeds. Water regularly and the roots will appear between the pieces of tape. Plant the whole tube into the ground or a large pot and watch the life cycle unfold.

Heart to Heart

Starting Points

- Act the part of a person who does all the wrong things to stay healthy. Ask groups to draw a large picture of such a person and to write around it all the things they are doing wrong.

- Talk about how to keep the different parts of the body healthy, particularly focusing on the heart. Discuss what the heart does and why having a healthy lifestyle is so important.

Making the Display

- Make a collage of profiles of two heads talking – having a heart to heart! Place the title 'heart to heart' between them.

- Place a large cross on one side and a tick on the other to represent things that are good for the heart and those that are not.

- Make large red hearts from collage materials and paper, and turn them into people, each with jointed arms and legs, and showing a fact about health and hearts, for example smoking hearts, lazy hearts and hearts with a poor diet. Place these under the cross.

- Make large red hearts with a healthy lifestyle and place under the tick. Discuss what these should be.

- Add speech bubbles to explain what each heart is doing.

Focus of Learning

Learning about the features of a healthy lifestyle
Understanding the purpose of the heart and how to keep it healthy

Further Activities

- Create a large picture of a real heart. Label the main chambers and blood vessels. Find out how it works and its function. Sew fabric hearts and label them with healthy activities. Arrange around the central heart.

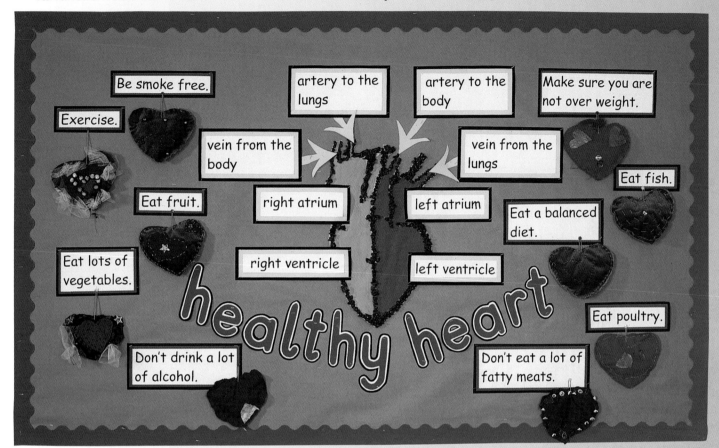

- Make up a series of questions and interview people to find out about their lifestyles. Choose one person and discuss how they might improve what they do to keep a healthy heart.

- Set health targets. Provide a piece of dowel as the basis for an arrow. Make a point from card. Decorate the dowel with feathers, wool, fabrics, ribbons and so on. Write a healthy heart target on a piece of paper and pierce it with the dowel arrow. Stick the arrows into a large heart made from red playdough. Set a date by which the target is to be reviewed and say how well the children have met their targets.

- Make up a food wheel to show the balance of foods we should eat in a day.

- Look at school lunches – either lunch boxes or the school's cooked lunches. How healthy are they? Which foods should there be more of and which less? Put together an example of a healthy lunch box.

- Hold a debate about smokers versus non-smokers, in which each is able to make their point. Take a vote at the end – to smoke or not to smoke!

Wriggly Worms

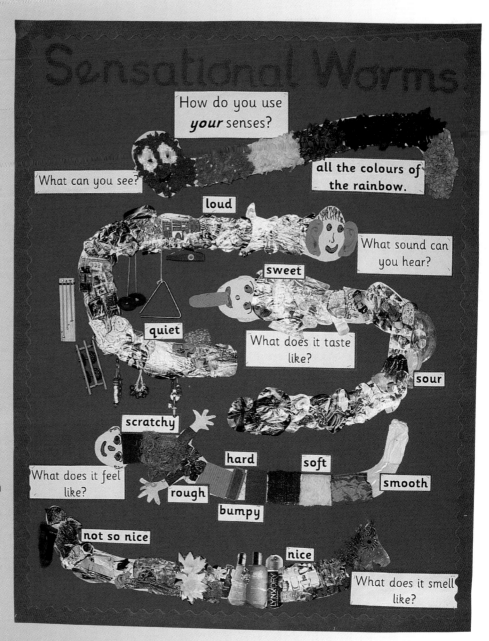

Starting Points

- Sing songs about the five senses. Point to eyes, ears, nose and mouth and wiggle fingers as each sense is mentioned.

- Ask groups of children to find something they really like to represent one of the senses. Make collections of things for each of the five senses.

- Introduce sensory worms, each of which has a special sense that it uses a lot. Organise children into groups and allocate a sensory worm to each group. Ask them to describe it, which part of the body enables the worm to use its specific sense and what sorts of things it does in its everyday life.

Making the Display

- Collect together collage materials that can represent sound, touch, sight, taste and smell.

- Use the materials to make collages of worms. Each worm illustrates one of the senses. A hearing worm could be made of tinfoil, crackling paper or plastic; a feely worm could be made of a variety of different textures and a seeing worm of different colours and patterns made using different techniques. The smelling and tasting worms are more difficult: cut out and paste pictures of different foods, spices and herbs and smelly plants on to the worm shape. For smells include other things as well as foods.

- Give each worm an exaggerated characteristic, such as a long tongue for the tasting worm, big ears for the hearing worm and so on.

- Mount the large worms on a bright background, arranging them so that they fill up the spaces. Label with the appropriate experiences.

- Add questions to do with the senses.

Further Activities

- Make hand puppets for a senses game. Give out the puppets to different children. Hold up an object that could be explored by one or more of our senses. The child wearing the correct hand puppet comes out front. For example, if a small cake is shown, then the children with the taste, smell and sight puppets come out. Talk about which of our senses we use the most often. What would it be like if we didn't have that sense?

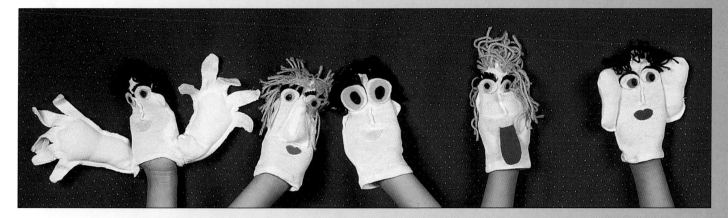

- Take children outside to explore some of the senses:

Sight Give pairs of children a card with different colours from paint samples stuck in a strip down one side of double-sided tape. Match the colours with natural things found outside and stick them opposite the samples.

Smell Provide a cup of coloured water for pairs of children and ask them to make it into a smelly mixture, using things they find.

 The children must not drink this mixture and must beware of poisonous plants.

Hearing Practise listening outside the classroom. What sounds can be heard? Which are loud and which are quiet? Sit outside and listen carefully to all the sounds. Which ones are nice/not nice? Which are the loudest? Which are the quietest? Put them in order from loudest to quietest.

We matched these colours with things we found

Taste Have a picnic together. Which things do the children like to eat? Which things do they not like? Ask the children to imagine they are a worm; would they like the same things?

- Decorate some plastic cups with a numbered worm and place some interesting smelly things inside each. Put fabric over the top. Add pictorial cards of the objects in the pots. Match the smell to the pot.

Snail Trail

Starting Points

- Go on a snail hunt around the school. This is best done on a damp day. Ensure that the snails are kept safely. An ideal home is a plastic aquarium containing the stones and plants on which the snails were found.

- Look at the snails in detail, the patterns on their bodies and shells, and name the different parts of the body. Use a hand lens. Draw and colour the different patterns carefully.

- Talk about the life cycle of a snail and create a leaflet about it.

Focus of Learning

Making close observations
Learning about the lifestyle of a familiar animal

Making the Display

- Make a border and background of leaf prints. Ensure that there are different shapes, sizes and shades of green.

- Ask each child to make and colour a large snail with a coiled shell. The snails need not be accurate, but should represent the basic shapes, colours and sizes in relation to the background. Cut round the coil so that it is free. Attach the shells to the display with a foam body for the snail.

- Add other details to the display using collage materials such as a wall, flowers, grass and plant pots.

- Use white paint to show the snail trail of each snail through the leaves.

- Add questions devised by the children, ensuring that some allow for investigation and others research.

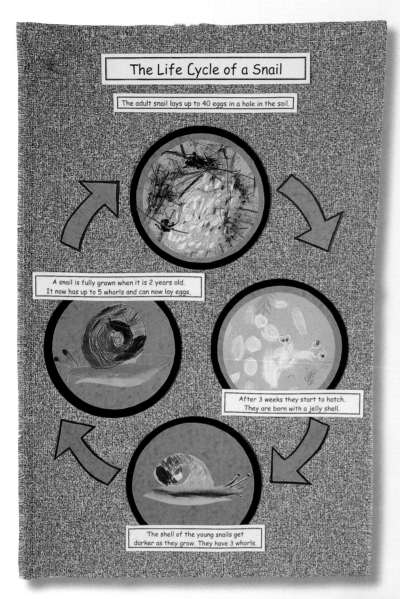

The Life Cycle of a Snail

The adult snail lays up to 40 eggs in a hole in the soil.

A snail is fully grown when it is 2 years old. It now has up to 5 whorls and can now lay eggs.

After 3 weeks they start to hatch. They are born with a jelly shell.

The shell of the young snails get darker as they grow. They have 3 whorls.

Further Activities

- Make a poster from collage materials to show the life cycle of a snail.

- Place live snails on black paper and observe the trail they leave. What makes the trail? Do snails always leave a trail? Try other colours of paper. On which does it show up best? Always return snails to the wild afterwards and talk about handling creatures carefully.

- Can snails smell? Place a snail in the middle of a sheet of paper and place some things around it, including only one smelly food. Does the snail go towards or away from the smell? Does it do this more than once? Try other smells. Discuss the results. If the snail did go towards the smelly food, how can we be sure it was smell? Could it have been the colour that attracted it? Try another test using similar colours and one smell.

⚠ **Do not include foods with a lot of salt, as these can kill snails.**

- Compare two minibeasts, using good photographs, for example a bee and a fly; a ladybird and a woodlouse. In what ways are they the same? In what ways are they different?

- Ask questions about one of the animals. Encourage the children to begin their questions with different words, such as what, how, when or which. Children often overuse the 'why' questions. 'Why' questions cannot be investigated, requiring a secondary source to find the answers.

- Provide groups with different shells. Ask them to describe the shells and what they think the animal that lived inside was like. Give them some modelling clay/plasticine to model what they think the animal might have looked like. Make observational drawings of the shells.

Drawings of our shells

What do you think lived in each of these shells?

Sensible Susan and Silly Samantha

Focus of Learning

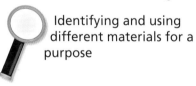
Identifying and using different materials for a purpose

Starting Points

- Go on a hunt around the school or classroom to name different materials. Record the names and draw the materials.

- Collect some objects made of the different materials and ask the children to sort them according to what they are made of. How could objects made of more than one material be sorted?

- Talk about the materials and what they are used for and why. Introduce the characters of Sensible Susan and Silly Samantha and their houses. Sensible Susan chooses all the right materials but Silly Samantha uses silly materials. Ask the children to suggest suitable materials for each house.

Making the Display

- Prepare a large display board with a green and blue background. Construct two large basic house shapes from strong but light cardboard. Decorate the walls for each house, using the chosen materials. It is not possible to use the actual materials for the sensible house, but they could be represented by, for example, sponging a pattern for bricks. Stick the houses to the wall using strong glue.

- Complete the background: the story could be taken further by creating birds made from silly materials, such as a body made of paper and wings made of clothes pegs. Sensible Susan's should be realistic.

- Label the parts of each house with the names of the materials. Ask the children to explain why each material is good or not so good for the parts of the houses.

Further Activities

- Explore waterproof materials. Test a variety of materials using droppers and decide which are the most water-resistant and which will let the water through.

- Play a 'materials' game. Split the children into pairs, and ask them to stand back to back. Give each child an object to hold and ask them to take turns to describe it to their partner, who draws a picture of it from the description and then identifies it. The name of the material(s) should be included in the descriptions.

- Design a shopping bag. Look at a collection of carrier bags. Discuss what they are made of and why. Talk about why they need to be waterproof. Look at their shape and handles. Make bags for the two characters. Test the bags to find out if they are waterproof and how much shopping they will hold.

- Gather a collection of building materials for children to explore and to name the materials. Read the 'Three Little Pigs' and ask the children to make up their own version of the story. What materials would they use for a house that would stand up against the wolf?

- Using what the children have learnt about water-proofing, ask them to design and make a hat for Sensible Susan that will keep the rain out and one for Silly Samantha that will not. Test the designs outside using watering cans.

Hot and Cold

Starting Points

- Look at a candle burning. Talk about safety first: ensure that the candle is in a tray of sand so that it is stable. Long hair should be tied back and the children should observe from a safe distance. Talk about fire and its dangers, and that the children should never play with fire. Make a warning poster.

- What is the candle made of? What other things are made of wax? Ask the children to observe and describe the flame, the different parts of the candle and how it changes as it burns.

- Show other familiar materials. Name each one and invite the children to say which they think will burn, melt or remain unchanged when heated up in a fire. Record predictions and test. Compare the results with the predictions. Look at secondary sources to show that glass and metals melt at very high temperatures.

Focus of Learning

Understanding how different materials change when they are heated and cooled

Making the Display

- On a blue background make flames for a fire, using red, yellow and orange twisted papers, feathers, string and so on. Represent smoke using cotton wool and grey paint

- Cut out silhouettes of people in the front of the fire. Ensure that these are big enough to be seen. Make other people shapes and dress in coloured clothes. Place these at the side of the fire.

- Place actual samples of different familiar materials in the fire – wood, paper, metals, plastic, fabrics, glass, rock, leather and so on. Name each one.

- Place the title above the bonfire and add focus questions about burning.

Candle Observations

When wax melts it looks like water.

The flame is all around the wick.

When the candle is burning there is liquid wax around the wick.

The flame flashes, flickers and turns.

As it burns the candle gets shorter and shorter.

When the wick bends over the flame becomes wider and higher.

You can smell a candle burning.

Further Activities

- Make a group candle display. Decorate different-sized rectangles of paper with wax-resist patterns. Cover with a thin paint wash. Curve the rectangles into candle shapes and stick on a background. Arrange the candles together and write observations of a burning candle on the displays using scientific language.

- Make a collection of liquids, such as tomato sauce, milk, fizzy drink, water, washing-up liquid, oil. Discuss that they are liquids because they can be poured; they flow. What other liquids do the children know? Is syrup a liquid? Is sand a liquid? Vote and discuss reasons for and against. Make up a list of rules for a liquid.

- Look at the collection of liquids from the previous activity. Discuss what the children think will happen to each of the liquids in the freezer. Test the predictions at school or at home and report back. Oily liquids, such as tomato sauce and salad cream, will not freeze in an ordinary freezer.

Our 'Freezing of Materials' experiments.

✓ = freezes   ✗ = does not freeze

	✗		✓		✓		✓
	Tomato Sauce		Milk		Mint Sauce		Comfort
	Baby Oil		Liquid Soap		Salad Cream	Water	Vinegar	
	✗		✓		✗	✓	✓ish!	

Our lollies

cola

orange squash

strawberry milkshake

cranberry juice

mint sauce

yogurt

milk

- Make ice lollies of different flavours. Select liquids and pour into ice trays. Add a lolly stick or a toothpick. Invite children to make up a new flavour, such as mint sauce. Can the children make a rainbow lolly? This may be done by using different colours, and partially freezing one before adding the next.

- How long does it take an ice lolly to melt? Link with Maths and time them melting. Do big lollies take longer than small ones?

Sandy Beach Holiday

Starting Points

- Go on a rock hunt in the local area to find out places where rocks are used, for example in buildings and other manufactured things, such as gravestones and statues.

- Use hand lenses to make close and detailed observations of a collection of six to eight different samples of rock. Include samples such as sandstone, limestone, marble, chalk, granite, pumice and slate. Can the children decide on a criterion for sorting them?

- Look at pictures of a beach or visit a beach to see sand and pebbles, rock pools and cliffs. Describe what is there naturally and what is manufactured.

Focus of Learning

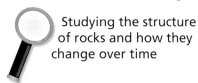 Studying the structure of rocks and how they change over time

Making the Display

- Set up the display board with a table beneath. Make a border for the board of shells or pictures of sea animals and seaweeds.

  - Make a background of sandstone cliffs and a sandy beach, using sandpaper and screwed-up painted paper or covered boxes. Cover the table to look like a sandy beach – include some sand, pebbles, shells and so on.

  - Paint plants and animals. Make seagulls and other birds. Hang some in front of the display.

  - Make sandcastles either from sand or sandpaper stuck on to card with glue.

  - Add painted people to the display, showing beach activities. Bend the figures so that the people look as though they are active.

  - Complete the display by adding labels and questions to do with rocks and the rock cycle.

Further Activities

- Observe the structure of real sandstone through a strong hand lens. Discuss how it is made. Make some 'edible' sandstone by binding breakfast cereal with chocolate. Compare with the real sandstone. Create a 'rock' that has different components in size and materials, such as a conglomerate, by adding raisins and so on to the mixture.

- Explore the rock cycle. Break up some sandstone to show how it is weathered to make sand. Mix sand with a little wet clay to show how it can be cemented to form sandstone. Talk about how this happens on the beach. How hard is rock? Test four different types of rock to find out how hard they are by asking the children to scratch the surfaces with a fingernail and harder objects, such as the handle of a fork or a nail. Can any of the rocks make a scratch on the others?

We made this sandstone using breakfast cereal and chocolate glue.

Nature made this sandstone from sand grains and clay glue.

- Compare the structure of granite and sandstone by building models to show the differences. Use Lego, preferably of different sizes and colours, to show how granite grains fit together and use marbles to represent the structure of sandstone.

Granite

Sandstone

- Which rocks are permeable? Pour water on to the Lego model (granite) and on to the marbles (sandstone). Which rock lets the water through? Test real rocks to find out if the result is the same. Granite is impermeable because it doesn't let the water through. Find other rocks that are impermeable and some that are permeable.

A Cup of Tea

Starting Points

- Have a tea party. Make a cup of tea as the children watch and list the ingredients. Which of the materials dissolve in the water and which do not?

- Discuss dissolving. What do the children think happens? Is it a reversible change? How could this be proved? Ask the children to use speech bubbles to record and share their ideas with the rest of the class.

- Ask the children to predict, citing reasons, whether there is a link between the size of sugar grains and the rate of dissolving. Record ideas and ask the children to plan how to test them.

Making the Display

- Back a display board in coloured paper and add a border.

- Make models or paintings of children's heads or take photographs of the children in the class. Arrange the heads on the board, leaving room for a speech bubble next to each head.

- Cut speech bubbles from paper to show the children's ideas about dissolving after they have completed the Further Activities on page 35.

Focus of Learning — Examining materials that dissolve and the conditions that change the speed of dissolving

Further Activities

- Create a poster to show how to make a cup of tea using scientific language. Include vocabulary such as dissolve, soluble, insoluble, mix, melt, temperature, evaporate, condense, liquid, solid, gas.

- Experiment to find out what might make a difference to how quickly a material dissolves. Grain size might make a difference to the rate of dissolving, but ask for other ideas, such as temperature and volume of water, mass, colour and texture of sugar and how many times it is stirred.

- Design and make tea strainers. Try out the tea strainers to find out which are the most effective. Test and evaluate each strainer.

- Design and make a teabag that will successfully survive immersion in hot water. Examine and compare the materials used to make them. How should the edges be joined so that they are safe to use? Try staples, different glues, sewing and tying with thread, plus any other ideas the children might have.

⚠ **An adult must supervise the use of hot water.**

Dissolving is a reversible change.

How to make a lovely cup of tea.

1. Heat the water in the kettle until it boils.

The bubbles are full of water vapour which is a gas.

2. Put loose tea into the teapot. Add the water.

The tea leaves are insoluble. The colour and flavour dissolve in the boiling water to make a solution.

3. Pour the tea into the cup through a tea strainer.

The tea strainer filters it. The liquid goes through the holes but solid tea leaves stay in the tea strainer.

4. Add the milk to the tea.

The milk cools it down and dilutes it. Water evaporates from the surface.

5. Add sugar and stir.

Sugar dissolves in the tea.

A lovely cup of tea!

Our teabags

Keeping Warm

Focus of Learning

Proving that some materials are better insulators than others
Understanding that temperature is a measurement of how hot and cold things are

Making the Display

- Represent the inside of a takeaway shop, for example a fish and chip shop, Chinese takeaway, or hamburger van.

- Construct a shelf across the front of the display and place on it containers made from the following materials: foil, cardboard, polystyrene, clear plastic and newspaper.

- Label each container with the name of the materials, and place 3D models of samples of takeaway food in them.

- Make the back of the shopkeeper's head using collage materials and attach this to the front of the shelf, as if facing the customers. Add some customers.

- Attach a title to be investigated.

Starting Points

- Look at the packaging used for takeaway food. What is it made from? Why?

- Ask the children to bring in clean takeaway packaging to school, for example card, polystyrene and other plastics, different papers, metal and so on.

- Challenge the children to make a container that will keep water warm. Empty drinks cans make useful metal containers, or use plastic paint pots. Wrap each can in a different material and take the temperature at regular intervals. Record on a table and transfer to a graph. If possible, use an IT package and a digital thermometer.

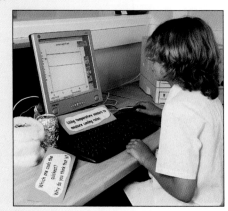

Further Activities

- Make thermometers, using small water bottles, coloured water and a narrow straw. Fill the bottle up completely with the coloured water; place the straw in the neck and plug around it with plasticine so that no water can escape. Place the bottle in cold, warm and hotter water and observe what happens to the level of water in the straw. Water expands on heating, so the water will rise up the straw as it gets hotter.

- Carefully observe the scale on an alcohol thermometer. Teach the children to read the thermometer accurately and note how it changes in warm and cold places. Compare the temperatures in different places around the school. Where is the warmest place? Where is the coldest?

 Mercury thermometers are not appropriate for use by young children.

- Put the round end of spoons made of metal, plastic and wood into warm water. Feel the handles of each and explain what is happening to the temperature. Straighter-handled spoons are better, as they allow greater conductivity of heat up the handle.

- Make an ice hand by filling a plastic glove with water and freezing it. Ask the class for ideas about how to make the hand melt more slowly. Try wrapping the ice hand in different materials. Compare the results. Talk about how good insulators slow down the movement of heat from a warmer to a colder place and vice versa.

A Box of Gas

Starting Points

- Show the children examples of water in its three states – ice, liquid water and water vapour – and demonstrate the changes when ice is heated from one state to another. Compare and discuss the three states. List the differences between a solid, liquid and gas.

- Sort a collection of materials into solid, liquid and gas (see chart on p.39).

- Focus on different gases. Provide objects containing gases, such as an air-filled and a helium balloon, a fizzy drink, a bag of crisps and a light bulb. What is inside each of them? (Crisps are packed in nitrogen.)

Focus of Learning

Learning the names and characteristics of familiar gases

Making the Display

- Make zig-zag books for each gas explaining their uses and properties.

- Research 'gas facts' about the more common gases: helium, oxygen, carbon dioxide, carbon monoxide and nitrogen. Others may be included, such as hydrogen (interesting but explosive!). Provide access to a range of secondary sources, including the Internet.

- Make a large box for each gas and decorate it to show information about where the gas is found in our everyday lives, for example carbon dioxide is taken in by plants and given out when we breathe, when things burn and by fizzy tablets when water is added.

- Place objects inside each box to do with the gas. Invite others to add to each box.

- Display the boxes together with large labels showing the names of each gas.

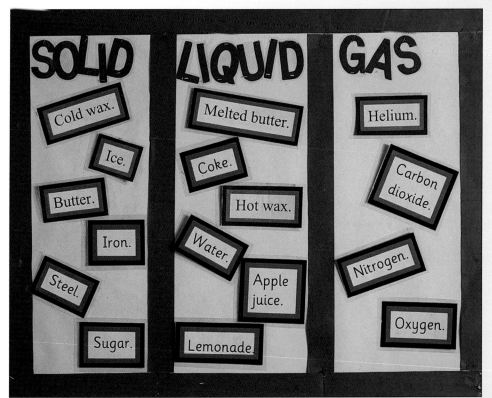

- Compare a helium balloon and an air-filled balloon. Why does the helium balloon rise while the other does not? (Helium is lighter than the surrounding air, so rises; expired air inside a balloon contains more carbon dioxide than the surrounding air so is heavier and it will sink.) Can the helium balloon be made to balance in mid-air?

- How heavy is gas? Weigh a cup of fizzy drink on a digital balance that measures small changes. Compare the weight over a period of time. What happens? After a week, taste the drink. How is it different? Ask the children to explain the change in weight.

- Place fresh, living yeast in a small container. Add sugar and some warm (not hot) water. Place a balloon over the neck of the container and watch what happens. Carbon dioxide is produced, which inflates the balloon. Explore what happens under different conditions. Is gas produced if there is no sugar or the water is cold?

- Make a volcano from a pile of sand or papier-mâché. Place a plastic container in the crater. Put baking powder inside and add vinegar. Watch the volcano erupt! Add food colouring for more spectacular results. What other mixtures can children suggest to try? Discuss with them what gas is made and that this is an irreversible change.

Litter Bugs

Focus of Learning

Learning what materials provide litter Disposing of and recycling waste materials

Starting Points

- Look at the places in the school where there is the most litter and try to explain why it is worst there. How could the school be made a litter-free zone?

- Decide as a class how to improve the litter problem in an area – it could be in the school or outside. Write letters to people or newspapers, make posters, make up questionnaires and so on. Collect information on how well the campaign works.

- Collect different items of clean litter/packaging and sort them into the materials they are made of. Which materials are used most?

Making the Display

- Make a background for the display board using things made from the found materials, for example a street light, a car and so on. Show a litter bin containing rubbish.

- Make models of litterbugs from as many different materials as possible, such as paper, card, metal, plastics and wood. Show them doing something such as eating a piece of rubbish, sweeping it up or putting it in a bin. They should be varied in size and type.

- Label each model with its name, chosen to represent a particular characteristic. Mount some models on the background and suspend some in front of the display.

Further Activities

- Make some recycled paper from classroom waste. You will need: waste paper (finer papers, such as tissue paper, are easier to recycle); a large bowl of warm water; an eggbeater; a wire gauze or a fine mesh (for the size of paper to be made); a teaspoon of instant starch; sheets of absorbent paper (blotting paper or newspaper). To make the paper: tear the waste paper into small pieces and put it in a large container. Add plenty of warm water. Add the starch and leave the mixture for ten minutes and then beat it to a pulp

with the eggbeater. It must be really mushy. Leave the pulp for several hours or overnight. Dip the mesh into the pulp so that the pulp lays flat on it. Let the water drip out for some minutes. Turn it upside down on to the absorbent paper. Cover with more absorbent paper and leave overnight. Remove the absorbent paper and let the recycled paper dry out completely. How good is the paper for writing on? How could it be made better? How is the recycled paper different from other paper?

- Talk about waste materials that can be recycled. How are people encouraged to recycle materials in the area?

- Discuss plastics and find out how they are made. Make a collection of the many different types and list their uses. Are plastics natural or manufactured materials?

- Collect some empty plastic bottles. Think of as many uses as possible for them. Make them and share ideas.

- Design litter bins for inside at school – use large supermarket boxes to represent creatures with open mouths or paint interesting patterns and designs.

The Wizard's Kitchen

Focus of Learning

Recognising and classifying changes as reversible or irreversible

Starting Points

- Talk about reversible and irreversible changes and name some examples in the environment.

- Carry out a series of activities in which the children explore changing materials and decide whether the change is reversible or not. Include: mix sand with water; mix salt with water; melt ice; make toast (observe safety rules); make porridge (observe safety rules); let a painting dry; put a fizzy tablet in water.

Making the Display

- Back a display board in dark colours. Create a sorcerer and add this to the display.

- Show snow falling with collage materials (and show it melting in the background to illustrate freezing and melting).

- Make articles and contents for jars in the laboratory – use flubber, slime, playdough, plastic and Plaster of Paris, all of which illustrate irreversible changes (see Further Activities for recipes).

- Make a book containing 'spells' written by the children on charred paper as an example of burning.

- Illustrate reversible changes by adding packets of powders on shelves, some of which are soluble in water and others that are not. Be careful when using these. Washing soda is not recommended for use in primary classrooms.

- Make candles (see Further Activities) and place these on a table under the display board.

- Add questions about reversible and irreversible changes.

Further Activities

- To make flubber (irreversible change): make two solutions (A and B). A: mix together ½ cup of warm water, 1½ teaspoons of borax and food colouring. B: mix ¾ cup of warm water and 1 cup of PVA glue. Combine A and B in a mixing bowl. Note how the mixture changes. Knead it and store in a bag.

- To make slime/gloop (irreversible change): mix one part liquid starch and two parts glue with a few drops of food colouring. It will be really slimy!

- To make plastic (irreversible change): heat ½ to one cup of double cream until quite hot, add 2–3 teaspoons of vinegar drop by drop, continually stirring. Stir until the mixture becomes rubbery. Cool and wash when solid. Mould into eyeball shapes for the display.

 The children must be supervised at all times when using a cooker and must not eat the eyeballs.

- To make candles (reversible change): heat up a large pan of water. Place wax/coloured crayons in a smaller metal container and put into the pan of water. Allow the wax to melt. Give each child a piece of wick. Try different threads and strings. Arrange the class in a circle and allow each child in turn to place the wick in the hot wax. They then join the end of the line. This gives the wax time to solidify before the child can repeat the process. In this way, the wax builds up around the wick.

 The children must be supervised at all times when using hot water and melting wax.

- To make playdough (irreversible change): the basic recipe is one cup each of flour and warm water, 2 teaspoons of cream of tartar, 1 teaspoon of oil, ¼ cup of salt and food colouring. Add all the ingredients together, adding the food colouring last. Stir over a medium heat and knead until smooth. Place in a plastic bag to store.

- To make Plaster of Paris fossils (irreversible change): grease a wooden board and place a leaf on it, vein side up. Make a wall of plasticine around it and pour in Plaster of Paris. Allow to set. Turn over and remove the leaf to see the fossil.

- The Plaster of Paris could also be used to make candle holders. Pour the Plaster of Paris into a cardboard cuff on a greased wooden board. Place a pencil into the centre and allow to set. Remove the pencil when set. Decorate the holders with paint.

Washing Day

In the display photo, labels read: washing day — fur fabric — chiffon — fleece — cotton — rayon — linen — 'Observe my fabrics' — Which fabrics are absorbent? — Which are waterproof? — What happens to the water from the wet fabrics as they dry? — In what type of weather does washing dry the quickest? — **In what order will these fabrics dry?**

Starting Points

- Show the children a variety of fabrics and other materials. Which ones do they think are absorbent? Which ones are waterproof?

- Select a variety of fabrics of a similar colour. The fabrics will obviously have a variety of other properties – thick and thin, different textures, different types of weave and so on. Find as many similarities and differences as possible. Use hand lenses to examine how the fabrics have been made (woven, knitted or felted).

- Select another property for a collection of fabrics, for example thickness or what they are made from, for further comparison.

- Choose an absorbent paper and find ways to make it waterproof. Use soap, oil, candle wax, crayon and other materials that the children choose.

Focus of Learning

Establishing that some materials are more absorbent than others
Observing the factors that affect the speed of evaporation

How can we make the absorbent paper waterproof?

Making the Display

- Make up a washing line for the fabrics already explored.

- Mount on a sky background.

- Cut out the fabrics to the same shape and size.

- Soak each one in watered-down PVA glue. Alter the shape as though each is blowing in the wind. Allow to dry.

- Peg the fabrics to the washing line and add questions about drying.

- Add a washing basket with further fabrics that the children may use in their investigation.

Further Activities

- Do all liquids evaporate? Do they all evaporate at the same rate? For speed, use wide, shallow dishes and a small amount of liquid or, for accuracy, use more liquid and beakers. The containers used must allow for easy observation/measurement of the amount of liquid left.

- Test papers/fabrics for absorbency:

  - Cut out strips of at least four different fabrics. Predict how absorbent each one will be. Hang the strips over a tray of coloured water, ensuring that the ends are dipped in. Observe at regular times and keep a record. Ask groups of children to select how they could best record what happens.

  - Drop a measured amount of water on to identically sized squares of different papers/fabrics, measuring the area that becomes wet.

  - Compare the mass of fabrics before and after soaking with water.

- Find everyday examples of evaporation, for example watercolour paint drying, puddles getting smaller, water in the cat's bowl evaporating.

- Set up a problem-solving activity. For example, Ali needs an apron to keep him dry when he is washing up in his café. He will throw it away at the end of the day. Ask the children to design an apron for him, using the most suitable materials. Model these at the end of the lesson.

Circus Circuits

Focus of Learning

Making simple series circuits using batteries, bulbs, buzzers and motors. Discussing the dangers and usefulness of electricity

Starting Points

- Go on an electricity hunt at school to find everyday things that use electricity. Paste pictures from catalogues of electrical appliances on to a plan of a house to show where they are used.

- Ask groups to make a 2.5V or 3.5V bulb light up, using a 1.5V battery and wires. It is best not to provide holders for the bulbs at the start so that the children see where the wires need to touch the bulb to make it light up.

- Provide bulb holders and make the bulb light up. Ask the children to trace the path of the electricity and introduce the term 'circuit'. Exchange the bulb for a motor and then a buzzer. How can the motor be made to rotate in the opposite direction?

Making the Display

- Paint the background green and represent a circus tent using striped fabric or paper.

- Cut out two large clowns' heads. Paint them brightly, making one happy and the other sad.

- Make the nose of one clown by covering a transparent container with red paper and mounting a bulb in a holder inside it. Attach long wires to the holder and position on the clown with the wires behind the head.

- Cut out a bow tie; decorate it with paint, sequins and so on and mount on the spindle of a motor. Attach wires as for the nose. Attach the motor to the clown's neck so that the tie is free to turn.

- Place two batteries on a table underneath. Ask the children to attach the wires to each end of the battery to make the nose light up and the tie rotate.

- Add questions about electricity.

Further Activities

- Collect toys that use electricity to work. Trace the electric circuit in the toy. Sort the toys into sets: electricity makes them move; makes sound; makes them light up. Some toys may fit into more than one set. Link to work in Maths.

- Ask the children to imagine what it would be like at home if there were no electricity. What things would they not be able to do? Link with History.

- Discuss with the children that although electricity is very useful, it is also very dangerous. Emphasise that they should never put anything into mains sockets. Take time to ensure that they all understand this. Design a danger sign and an OK sign for electricity to put in the correct places in the classroom.

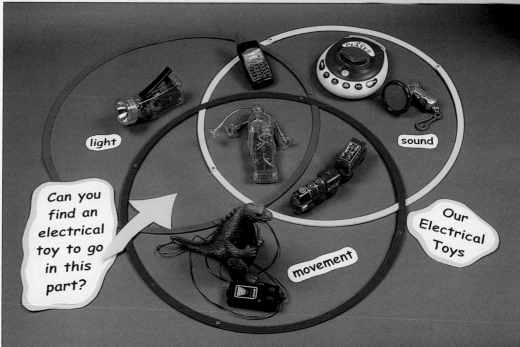

 Note that activities with children in the classroom should never use mains electricity, always 1.5V batteries or multiples of these.

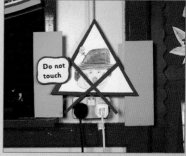

- Talk about where electricity comes from. Act out the story of how electricity gets from a power station to the home using some children dressed in bibs to represent the different parts of the circuit. Identify a child to be the power station who passes the electricity along a line to a second child who represents the house. This child passes the electricity on to various devices/ children, each of whom acts out the movement, makes the sound or shows how the device lights up when electricity is switched on.

Light and Dark

Focus of Learning

Comparing day and night

Finding out about animals that are active at different times

Starting Points

- Make the classroom dark. How do the children feel? Explore any fears concerning the dark. Ask the children what they might see if they were outside in the dark, leading to talk about nocturnal animals. Discuss the night sky, pointing out that the Moon can sometimes also be seen during the day.

- Discuss the environment under the ground, where it is dark. What animals live there?

- Make the classroom light again. Talk about animals that we see in the daytime.

Making the Display

- Split a display board into two. Complete the background using sponge paints.

- Put stars, the Sun and the Moon on the display. Discuss where to put the Moon on the display and what shape it should be.

- Attach dark netting over the night-time portion of the display to represent darkness.

- Complete the daytime scene of trees, made with layered papers of green leaves, flowers and clouds.

- Paint pictures of animals for each background and stick these on to the display; also attach some soft toy animals.

- Make mobiles of birds, bats and butterflies by cutting out symmetrical shapes. Decorate these using paint and suspend in front of the display.

The dark, dark, cave.

Further Activities

- Ask small groups or pairs of children to find out about one of the animals on the display. Ask them to share what they have found out with the rest of the class, or to record it in zig-zag books or on large sheets of paper.

- Explore living in the dark. Tell the story of 'The Three Bears'. Talk about where real bears live. Make a cave for a bear for its winter sleep. Use thick fabrics, cardboard and blankets. Pretend to be bears and think about how to make the cave light inside. Read the book *We're Going on a Bear Hunt* by Michael Rosen and Helen Oxenbury (published by Walker books) and act out the book using the bear's cave as the ending point for the rhyme.

- Find out about the things that give us light. Provide a range of familiar objects and ask the children to choose those that give us light. Include obvious things, such as a torch and a candle, but also a mirror. More able children could sequence the light sources from dimmest to brightest.

Sun

Which things give us light?

What shape is the Moon?

- Look at a picture of the Moon. Discuss whether it is 2D or 3D and sort a pile of objects into those that represent the shape of the Moon and those that don't. Observe the Moon (at home if necessary) on different days. When is the Moon in the sky? Is it ever in the sky in the daytime? Does the Moon change its shape?

Electric Elves

Starting Points

- Review what the children remember about making circuits by providing wires, a 2.5V or 3.5V bulb (not in a holder) and a 1.5V battery and asking them to light up the bulb.

- Explore the electrical conductivity of different materials. Organise the children into groups of three or four and ask them to set up a circuit using three wires, a battery and a bulb. Then make a gap in the circuit. Try putting different materials into the gap. Sort materials into those that will let electricity go through easily and those that will not. Introduce the terms electrical conductor and electrical insulator.

Focus of Learning

Identifying good and poor electrical conductors
Recognising how switches work to control circuits

Making the Display

- Make the background of card or paper but not foil. Paint the background for the chosen scene.

- Make the characters' bodies from metal pieces on a flat surface. Try to include a variety of metals, such as copper, brass, aluminium as well as iron and steel, but avoid any rusty metal. Ensure that all pieces of metal are touching in each body and that the arms and legs are movable.

- Mount the metal bodies on the background. Set up the circuit. Use three wires. Hide a battery in the display and attach it with one wire to a motor or light. A second wire leads from the bulb/motor to the hand of one character and the third from the battery to the hand of the other character. The hands act as a switch. When they are touching the circuit is complete.

- If the circuit does not work, there may be a loose connection or a stronger battery may be needed. If using more than one battery ensure that they are arranged with the negative terminal touching the positive of the next.

- Make the characters' heads of old tights, stuffed with foam or other soft materials, and attach. Add labels and questions.

Further Activities

- Examine how a real torch works. Trace the circuit. Make torches from cardboard tubes or ask the children to create an interesting design from other junk materials. Include a working circuit to make the torch light up.

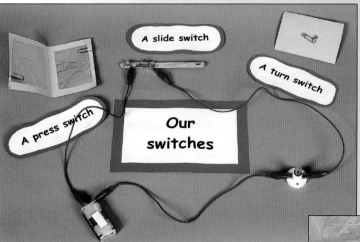

- Make switches, using knowledge of materials that are good conductors and those that are not. These may be turn, press, slide or tilt switches. Encourage the children to think of their own designs.

- Make up a 'human circuit game' to show that a complete circuit is needed for electricity to flow. Stand in a circle; when all hands are linked the electricity is able to flow and the device works. Devise sounds for the circuit and particular movements for bulbs, motors, buzzers, wires, battery and switches.

- Make an electric question game. Cut out a rigid rectangle from a cardboard box. Give the game a title. Write questions in one column and answers in another. Use loose cards, so that they can be changed for other contexts. Attach metal clips beside each question and answer. Make a circuit with a bulb, a battery and two loose wires. At the back attach a wire from each question clip to one of the answers, making sure that the correct answer matches the question clip. To play, attach one wire to a question and the other to the answer. If it is correct, then the bulb will light up!

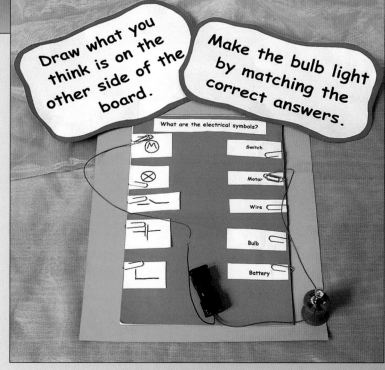

Disco Lights

Starting Points

- Shine torches on to transparent and opaque materials to see what sorts of shadows each makes.

- Try to make coloured shadows.

- Find out the children's ideas of where their shadow is on a sunny day. Fold an A4 sheet of paper in half. Ask them to draw a line halfway and to draw themselves standing on the line. Draw the Sun in the sky and then on the bottom half draw their shadow.

- Make individual transparent designs/objects using sweet papers and other transparent coloured materials. Do the same with opaque materials.

Focus of Learning

Recognising that light passes through transparent materials
Recognising that shadows are formed when light is blocked and are on the opposite side to the source of light

Making the Display

- Create a border in shiny paper and put on a title in bright colours.

- Cover a football or similar with shiny mosaics or shiny paper to represent a glitter ball (or use the real thing). Hang it in front of the board.

- Make three large figures of dancers in disco clothes. Stick each in different positions on the dance floor.

- Shine a torch directly above each one in the position of the glitter ball and draw round the shadows each one makes. A shadow directly below the light will be short. Those to the side are on opposite sides to the light source.

- Cut out silhouettes of each shadow and stick in the correct place. Add interest by allowing the shadows to come out of the frame.

- Complete with questions/statements.

Further Activities

- Select a simple shape and make different shadows with it. Cut out the different shadows. Add some other shadow shapes that are not of the object. Can the children decide which of the shadows are of the object and which are not?

- Look at a variety of papers and sort into categories of transparent, translucent and opaque. Make transparent and opaque characters/patterns. Use transparent materials to make patterns on a window. Find out how the number of layers of transparent materials affects how well the light can shine though them. Shine through one, then two, then three layers and so on until no more light can get through.

- Provide pieces of black, white and silver card. Predict which will make the darkest shadow. Test to find out.

- Set up a shadow stick outside on a sunny day. Use a stable object, such as half a lemonade bottle with a stick held firmly in it with sand. Plot the shadows an hour apart during the morning. Predict what will happen to the size and direction of the shadow in the afternoon. Continue to record at hourly intervals and compare results with the predictions. Discuss why shadows get shorter and then longer during the day and why they change position. Make a table and then a graph of the results. Ask the children to devise their own questions about the results and try to explain the differences in position and length. Record the shadow positions and length at another time of the year and compare the results. Link the difference in results to the position of the Earth in relation to the Sun.

- Ask the children to draw an object with its shadow but make a mistake, such as the shadow in the wrong position/with the wrong features/the wrong shape. Can others identify the mistake?

Blast Off!

Starting Points

- Ask the children to find as many different ways as possible to make air around them move and record these in a list.

- Hold a competition to move paper shapes a certain distance. Blow down straws to illustrate that the greater the force, the greater the movement and that force also has direction. Use scientific vocabulary: force, friction, direction and so on.

- Introduce rockets to the children by showing them a ready-made model (see Making the Display). Show how the rocket works by squeezing its sides.

- Discuss the forces acting on the model rocket at different stages of its flight. What force makes it come down to Earth again? What other forces are acting on it when it travels through the air? What are the directions of the main forces on it as it travels through the air? The main force on the rocket is gravity pulling it down towards the Earth. Air resistance also has a slowing effect.

Making the Display

- To make the launchers: use empty plastic washing-up liquid containers of the same volume and remove the tops. Attach a piece of cardboard tube to each top and decorate.

- To make the rockets: make cones from semi-circles of paper or thin card. Ensure that they fit on the top of the launchers with no air gaps. Make a selection of cones of different sizes.

- Stand the rockets against a background to represent space and add questions about forces to the display. Squeeze the sides of the launchers to make the rockets blast off!

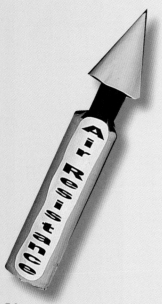

Further Activities

● Predict and compare how the size of the rocket affects how far it travels. Make a diagram of it travelling through the air and show the direction of gravity and air resistance on it.

Does the size of the rocket make a difference to how high it will go?

Air–powered rocket design number 2

Which rocket will travel the furthest? Why?

● Make another rocket design which works in the same way as the display model. Make the launcher as before, but with a wide straw and a plasticine plug at the neck. Make rockets from thinner straws with plasticine sealing one end. Place the rocket inside the neck and squeeze. Which length of rocket will travel furthest – the short, medium or long one?

● Make some air-powered buggies. Create the chassis of the buggy from thin card. Decorate this using collage materials. Push straws or dowel through the chassis and tape to the underneath so that they are fixed. Add cotton-reel wheels, making sure that they are able to move smoothly. Attach a small amount of plasticine to the end of each dowel to stop the reels falling off. Attach a piece of card with a hole in it for a balloon. Blow up the balloon and watch the buggy move! The lighter the buggy is, the further it moves.

● Create a jet balloon in the classroom. Blow up a balloon and let it go. Do it several times and trace the direction it moves each time. It will be haphazard. Talk about how the direction can be controlled. Blow up a long balloon. Keep it inflated with a bulldog clip. Attach the balloon to a wide length of straw with tape. Push a long length of thin, smooth thread through the straw and attach it to the classroom walls. Remove the bulldog clip and it will travel along the thread.

What might make a difference to how far they go?

Jet-propelled buggies

READY...STEADY...GO!

Spring Chicken

Focus of Learning

Learning that springs stretch and return to their original shape and exert a force on what is stretching them

Starting Points

- Show the children a spring. Where might it have come from?

- Show a pen with a spring inside. Discuss what the spring is for. What other things have springs?

- Challenge the children to find as many examples as they can at school and at home.

- Make some springs from card and paper, by pleating, folding and cutting into spirals.

Making the Display

- Make a plain green background with a border. Add grass made from curled paper at the bottom.

- Cut out letters for the title. Mount them on the background using circles of paper attached to the back so that the letters stand out and spring when pressed.

- Make a chicken shed from a large supermarket box and attach it to the wall.

- Sew together a large chicken from fabric or make it from paper and card. Add wings, beak, eyes and legs.

- Make a large spring from thin card to represent the chicken spring (it will not be springy). Attach this from the chicken shed to the large sewn chicken.

- Make small chicks from cotton wool balls painted yellow. Create some springy legs for them made by folding paper and also some zig-zag legs that will stretch. Add curled wings.

- Add labels to show the direction of the main forces acting on the chicken.

- Make some models with springs using various techniques. For example:

    – a jack in the box. This could be linked to nets in Maths

    – a box of spring flowers that will spring up when the box lid is opened. Make the stems from folded paper

    – creatures with stretchy arms and legs

    – springy frogs or bunnies that use elastic bands to make them jump up.

- Investigate elasticity. Make a collection of elastic bands and see how far each will stretch. Feel the pull backwards on the hand when they are stretched. Do the bands return to their original size and shape?

- Make ballistas using three pieces of cane, joined at the ends with elastic bands into a tetrahedron. Suspend a light cardboard or polystyrene cup in the centre using attached elastic bands. Project light objects by pulling back the cup and releasing. How far does the object move? What is the effect of pulling the cup back further or increasing the mass of the object? Ensure safety!

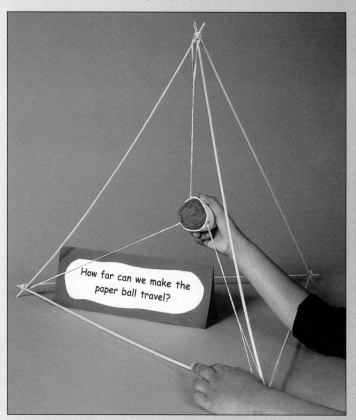

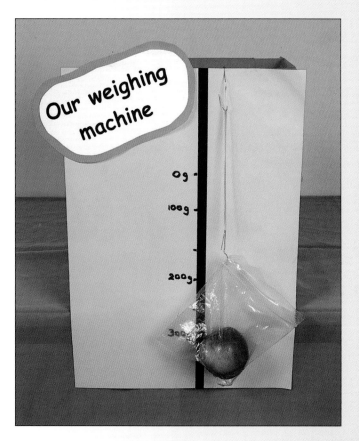

- Make an elastic-band weighing machine. Use a strong cardboard box weighted down with a brick inside. Attach a strip of paper to the front. Suspend a strong elastic band from a paper clip hooked over the top of the box. Attach a light bag as a holder for masses on the end of the band. Ensure it hangs freely. Add different masses and mark the amount of stretch each time a mass is added. Mark out a scale. Note that elastic bands do not stretch evenly. Use a stronger (wider) elastic band. How does this affect what can be weighed? Is another scale needed? Why?

Swing, Bounce and Stretch

Starting Points

- Find and label things in the classroom that are moved by pushing or pulling. Find different things in the playground that can swing, bounce or stretch. Which are pushes? Which are pulls? Ask the children to draw pictures of things they pull and things they push.

- Do a stretchy test together. Look at different things to find out which will stretch and bounce, for example elastic of different thickness, fabrics, dough, string and wool. Which are stretchy? Put them in order of stretchiness.

Focus of Learning

Experimenting with pushing and pulling movements

Making the Display

- Arrange a pole in the corner of the classroom, suspended between two walls so that the children can approach from either side. Ensure it is attached safely.

- Decorate the pole with a stretchy snake.

- Make 3D animals in small groups. Small boxes make good bases for heads and bodies. Turn them inside out to make painting easier and glue into the required shape.

- Suspend each animal on the pole. Attach a stretchy thread by cutting a slit in the top of the animal. Attach a matchstick or lolly stick to thin pieces of stretchy materials (pieces of old tights and so on), different elastics or non-stretchy materials. Put the matchstick into the slit and tie the other end to the pole.

- Add labels about stretching, bouncing, swinging, pulling and pushing to the display.

Which fish can swing?

Which can stretch?

Further Activities

- Create an underwater scene using a small round table. Suspend fish and sea creatures from the underside of the table. Decorate with fabric and seashells. Add questions about movement.

- Provide a range of fabrics and ask the children to find out if they stretch or not. How much will each stretch? Construct a chart to show the results.

- Make 'Stretchy Sid'. Create a body shape for a hanging person. Make the arms and legs stretchy by folding concertina style and attaching to the body.

- Make paper springs. Show the children how to draw and cut out a spiral from paper. Test the spirals to find out if they are springy or not. How could a better spring be made? Experiment with card, paper and wires in lots of different colours and sizes.

- Use the paper springs to make a springy animal. Cut out and decorate a simple animal shape. Attach it to the spring. Push down on the animal and watch it jump up! Challenge the children to adapt their springs to make the animal jump further. Alternatively, add more complex springs as arms and legs.

Which ones stretch?

How much does each one stretch?

- Test a collection of balls to find out which is the 'best bouncer'. Predict which ball will bounce best. How will we find out?

Mr Gumpy's Car

Focus of Learning

Learning that pushes and pulls can make things stop, start and change direction

Starting Points

- Read *Mr Gumpy's Motor Car* by John Burningham (published by Red Fox). Talk about how the car moves along and what makes it go.

- Look at model cars with moving parts. Which parts move? Which parts do we pull and which do we push?

- Discuss making a display to show Mr Gumpy's moving car. Decide on which moving parts could be included for the car and the background.

Making the Display

- Using a range of toy wheels, print the border. Make the background a road going through a country scene. Use a variety of techniques, such as sponging and adding papers, to produce effects for fields, the sky and the cobbled road. The leaves of trees could be rubbings of real leaves.

- Add things on the background that may be moved by pushing or pulling using a piece of attached card, for example the eyes of an animal, a plant, clouds that move across the sky or an animal that springs back when pushed.

- Make a 3D model of Mr Gumpy's car. Use stiff card for the side of the car. Ensure that some parts of the car can be moved by pushing or pulling, such as wheels, windscreen wipers, a door that opens, or a roof that pushes back.

- Collage a 3D Mr Gumpy to ride in the car. Attach the car to the display board.

- Add labels and questions about pulling and pushing.

60

Further Activities

- Design and make a moving wheeled model using dowel and junk materials. It is essential that either the axle or the wheels move but not both. Once the model moves freely, ask the children to predict how far it will go on one push. Discuss which one goes the furthest and why. Try changing the size of the wheels. Explore what happens to movement when the holes in the wheels are slightly off centre. There will be an 'up and down' motion, ideal for a clown's car!

- Push the models on different surfaces, rough and smooth. How does the surface affect movement? The children might find that the wheels slip on a very smooth surface, while on a very rough surface it is difficult to get moving.

- Look at real tyres and make prints of the treads. Talk about why a tread is necessary. Relate to the soles of training shoes. Wind elastic bands around the wheels of the models and see how this affects movement.

- Make push meters and use to measure how much push (force) is needed to shut a door, or make a toy car or brick start moving on different surfaces. To make the meter: cut an elastic band and tape each end to a cotton reel. Fit the reel over the end of a piece of dowel and tape the middle of the elastic band to the end of the dowel. Put a card wheel immediately under the cotton reel. Measure in centimetres from the point where the reel touches the dowel without being stretched. Colour each centimetre section a different colour. To measure the push, hold the cotton reel, place the end of the dowel against the object to be moved and push until it starts moving. The card wheel will move along the dowel and stay there, so the number of sections (cms) it has moved can easily be measured.

61

Day and Night

Text on the display:
- How do the Earth and Moon move around the Sun?
- What time of day is it at each place?
- Night-time is when part of the Earth is facing away from the Sun.
- The Sun, Earth and Moon are spherical.
- The Earth rotates once on its axis every 24 hours.
- SUN
- EARTH
- Daytime is when part of the Earth is facing the Sun.
- It takes the Earth 365 days to orbit the Sun.
- It takes 28 days for the Moon to orbit the Earth.
- DAY AND NIGHT

Focus of Learning

Learning how night and day are formed

Learning about the relative movements of the Earth, Sun and Moon

Starting Points

- Find out the children's ideas about the Earth, Sun and Moon by asking them to complete a concept map, using the words Sun, Moon, Earth, planet and star, and to make as many links as they can, explaining the links. Discuss their maps and any misconceptions, such as 'The Sun goes round the Earth'.

- Find out the children's ideas of how day and night occur. Ask the children to draw labelled diagrams, but do not correct them until further work is finished.

Making the Display

- Back a display board in dark blue or black.

- Create the Sun and the Earth from papier-mâché. Make the Sun white.

- Represent the Earth as a black object with landmasses in white. Indicate the direction of rotation with an arrow.

- Draw white arrows from the Sun to show rays of light towards the Earth.

- Identify three positions on the surface of the Earth where it is midnight, midday and dawn.

- Make three pictures to represent these and link to the three positions with arrows. Use appropriate colours and different techniques for each picture.

Further Activities

- Make a model to show how the Earth and Moon revolve around the Sun. Make a base from stiff card or plastic. Cut out a wide strip of thick card. Poke a piece of dowel through one end of it. Place a large polystyrene ball on the end of the dowel to represent the Sun. Poke the other end of the dowel through the end of the card strip. Cut another piece of dowel and repeat, using a smaller ball to represent the Earth. Attach this dowel to the other end of the strip. By moving the strip, the Earth will revolve around the Sun. Attach a smaller ball to a dowel and another strip, and mount another strip on top of the Earth dowel (the Moon). This will revolve around the Earth. Assemble as shown on the base.

- Make a set of true and false fact cards about the Earth, Sun and Moon. Ask the children to sort them.

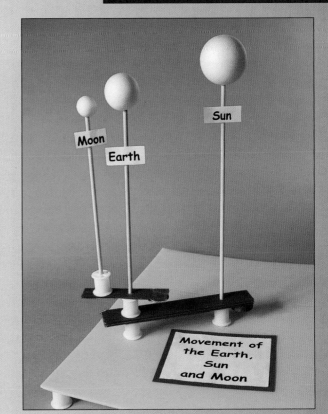

- Until man landed on the Moon, we did not know what was on the other side, as we always see the same face of the Moon from the Earth. Use two children to model how the Moon revolves around the Earth, always showing the same face to the Earth. Discuss with the children that as the Moon revolves around the Earth, it also revolves once around itself.

- Make some star constellations. Cut a small length of cardboard tube. Secure a piece of black paper over one end with an elastic band. With a pin, prick out a star constellation on the black paper. Hold the open end to the eye and point towards the light. The constellation shows up clearly. Make a selection for the children to look at over a period of time. Match the constellations to real ones in the sky or to a constellation chart.

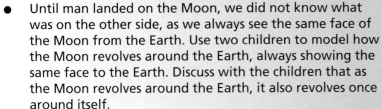

63

Changing Sounds

Starting Points

- Make a sound and ask the children to draw a diagram to show what happens to enable them to hear the sound. Compare their ideas.

- Show how sound travels in waves by using a Slinky toy. At one end give the Slinky a flick and watch the wave move along it. (This method is more accurate than portraying a wave with a skipping rope, as waves are made by causing the medium through which they are travelling to compress and release.)

- Discuss noise and noise pollution. What are the most annoying/loudest sounds? Which sounds are loud and which are quiet?

- Look at the structure of the ear. It is not necessary to look in detail, but the children should have an idea of the different parts and their importance in hearing and balance. Explain how loud noises over a period of time damage the ears, such as continued loud music.

Focus of Learning

Learning how we hear sound and how loud sounds can damage hearing Changing sounds in a variety of ways

Making the Display

- Make a large semicircle. Split it into six equal sections radiating from the bottom centre. Around the top mark the decibel ranges. Stick the semicircle to the background, leaving a space underneath for questions and pictures. Add the arrow marker.

- Label the semicircle so that it is clear which end of the scale is quiet and which is loud.

- Make two heads, one smiling and the other in pain. Ensure that the ears are clearly visible. Attach either side of the semicircle.

- Stick pictures of different objects that make a variety of sounds from quiet to loud in the correct section.

- Make a selection of other objects for the children to consider where each may be placed. Attach questions relating to sound.

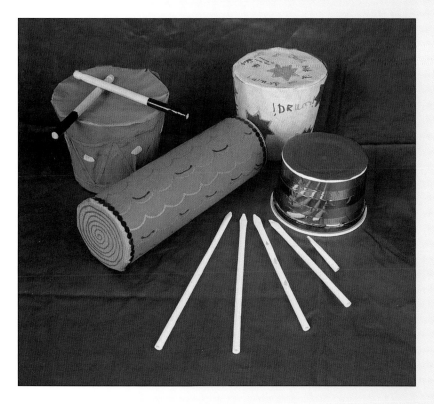

Further Activities

- Make pitched instruments. Create a straw pipe from a wide paper straw. Cut one end to a point and press the ends together so that there is only a small gap in-between. Place a long length in the mouth but don't let it get wet. Blow gently into it. It takes practice, but the children should be able to produce a low sound. Change the pitch by shortening the pipe. Make a generalised statement: 'The shorter the pipe, the higher the note.'

- Explore ways of changing loudness. Place thin paper over a comb and blow gently and harder; hit a drum gently and harder. Encourage the children to describe loudness as quiet or loud and not to use 'soft' (property of materials), or 'low' (could mean quiet or low in pitch).

- Find out how size affects pitch. Tap different-sized nails/bottles and bottles containing different amounts of water; pluck strings/elastic bands of different lengths stretched over an empty tissue box.

- Show how movement makes sound. Place rice or breakfast cereal on the drum and watch it jump when the drum is hit. Does it jump when there is no sound? When does it jump the highest? Introduce the term 'vibrate'. What vibrates when the drum is hit? Why does the vibration stop?

- Ask the children to design ear defenders. Which people wear ear defenders and why? Conduct a test to find out the best materials for ear defenders. Provide yogurt pots and a variety of materials to put inside or around them. Which materials keep out sound best? Decorate the ear defenders for a named work person, for example Rob the road digger or Diana the disco queen.

How We See

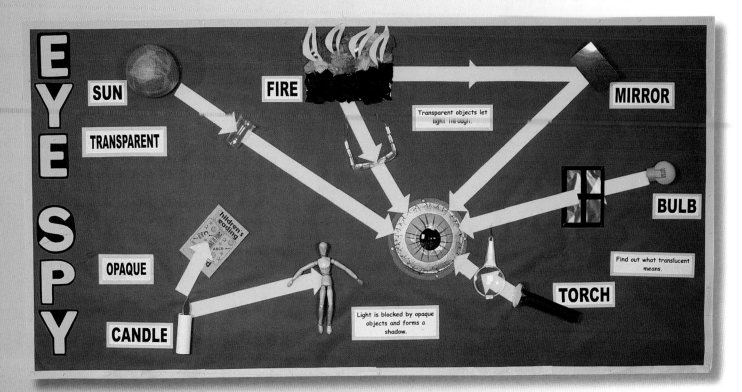

In the display: EYE SPY, SUN, TRANSPARENT, FIRE, MIRROR, OPAQUE, BULB, CANDLE, TORCH

Transparent objects let light through.

Find out what translucent means.

Light is blocked by opaque objects and forms a shadow.

Focus of Learning

Learning that light travels in straight lines and how we see light

Learning about the factors that affect shadows

Starting Points

- Make shadows from a variety of materials. Explain why some materials make dark shadows and others faint shadows. Discuss opaque, translucent and transparent materials.

- Hold a group discussion to define a shadow and a reflection. What is the difference? Explain to other groups using diagrams, writing and talking.

- Find out the children's ideas of how we see a light source. How many children think that light comes out of their eyes to see a candle flame? How many think that the light comes from the flame and enters their eyes? It is difficult to prove that the latter is correct. Explain and draw diagrams.

Making the Display

- Make the title 'Eye Spy' and place on a plain background.

- Make a large 3D eye from Plaster of Paris and stick it in the centre. Attach appropriate collage materials for each part of the eye.

- Make 3D models of different light sources and stick towards the edge of the board.

- Select some transparent and opaque materials or objects. Place one between each light source and the central eye.

- Add bright arrows to show the direction of light and how it passes through transparent materials and not through opaque.

- Add labels to the display.

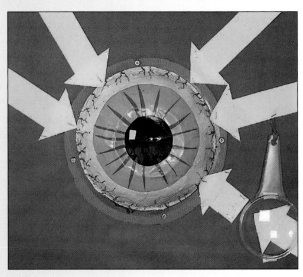

Further Activities

- Design and make photograms. These are black, white and grey pictures made on light-sensitive paper. The best method is to make them in a darkened space lit by a red safety light. The resources needed can be obtained from photography shops: light-sensitive paper; red safety light; a 15W source of light; developer in a tray; fixer in a tray; two trays of water to wash off developer and fixer; and tongs.

 Developer and fixer can irritate skins. The children should wear gloves and goggles and be supervised.

photograms

- Stand an anglepoise lamp with a 50W bulb about 70cm above a bench. Place a sheet of light-sensitive paper on the bench directly underneath it. Place an opaque object on the paper and a piece of clear Perspex or glass over it to keep it flat. Switch on the light and count to three. Using tongs place the paper in the developer. Ensure it is completely submerged and watch the picture appear. The object will appear white on a black background. When happy with the contrast, use tongs to place the paper in a dish of water and swish it about to wash off the developer. Now put the paper in the fixer and leave for about two minutes. This will ensure that the picture does not fade in ordinary light. Wash again thoroughly in the second water tray. Never mix the developer and fixer. Hang up on a line to dry.

- Find out how the eye works. It is not necessary for the children to know about its structure in detail, but to realise that light enters the eye, images are formed and nerves transmit messages to the brain.

- Make a simple periscope using lengths of plywood and mirrors. Fix together with wood glue. Show how the light is reflected from the mirrors so that we can use it to see things above us.

Can you make up your own secret messages?

Use the mirror to help you read them.

- Look at reflections and how they are different from the real thing. Place a mirror horizontally across a sheet of paper. Ask the children to write the alphabet on the paper. Which letters look the same in the mirror and which do not? Do the same with the mirror held vertically up the side of the page. Write some secret messages using mirror writing.

Sails and Movement

When the wind blows.

Area of sails

Stiffness of sails

Number of sails

Direction of the wind

Strength of the wind

Shape of sails

What might make a difference to how quickly the sails turn?

Starting Points

- Find pictures of things that have sails, such as boats and windmills. Discuss what the sails are for and what causes them to move. Observe how wind makes different things move, either outside or using a fan on a low setting. Find out about the Beaufort scale of measuring wind force.

- Make model windmills from squares of card. Make two diagonal folds in the paper to identify the central point. Cut along each diagonal fold to the point halfway to the middle. Put a pinhole at the centre. Take each section of a corner to the centre point and overlap. Push a headed pin through the centre. Push the back of the pin through a short section of plastic straw and then into a piece of dowel or a pencil. Blow on the sails and they should turn freely. If not, loosen the pin on the stick.

Focus of Learning

Learning that wind exerts a force to make things move
Proving that air resistance slows movement
Showing how surface area affects movement

Making the Display

- Make a background of land and sky. Keep it plain, using fabrics or paints.

- Create a large windmill from card. Look at the shapes of sails of real windmills for ideas.

- Attach the sails to the main windmill, using either sticky tape or a nail through the centre.

- Add model windmills of different sizes, with different numbers of sails, differently shaped sails and so on.

- Complete by adding the main question and the children's suggestions of what might make a difference to how the windmills move.

Does the area of the sail make a difference to how quickly the mass falls?

What is the force that slows it down?

What is the force that pulls it towards the ground?

Further Activities

- Does the surface area make a difference to how quickly paper falls? Drop pieces of paper with different surface areas, for example a flat piece of A4, one folded in half and one in quarters. Predict with the children how the paper will fall and test each one.

- Carry out an investigation to find out what makes a difference to how quickly the sails turn on a paper windmill. Different groups of children could investigate different aspects and relay their findings to other groups.

- Wind some thread with a small weight on the end around a cotton reel. Cut slots into one end of the reel. Fix a thin card sail in the slots. Ensure that the sail can move round freely on its axle. Let the weight drop. Using a stop clock, compare the falls of different sails.

- Create simple flat boat shapes from polystyrene trays. Make a mast from a short stick or cocktail stick and attach a sail. Create a wind using a small hand fan, or an empty washing-up liquid bottle can make a good blower when squeezed. The bottle blower works better if the top is removed and a wide plastic straw is inserted and held in place with a lump of plasticine. Squeeze it behind the boat. Investigate how the size and shape of a sail affects how quickly the boat moves in a track of water.

- Make a simple land yacht using dowel in a triangular shape, attached by elastic bands. Wheels can be empty drink cans or airflow balls. Add a sail and see how it moves in the wind.

Rosie's Yacht

What is the force that slows it down?

What are the main forces acting on the land yacht when it is moving?

Winnie's Sound Walk

Starting Points

- Read a story about a lot of different sounds, such as *Peace at Last* by Jill Murphy (published by Macmillan). Invite the children to make the different sounds in the story.

- Ask the children to make different sounds using only parts of their bodies. How many can they make with their hands/feet/tongues? Put three body sounds together to make a sequence. Find ways to represent different sounds, for example making a wavy line with a hand. Choose one of the children's ideas and make the sound while doing the action. Use other contrasting actions, for example sudden/slow; strong/gentle and so on. These could be a punch into the air, a sudden jerk of the head, a roll of the shoulders or head and so on. Make loud and quiet sounds, high and low sounds. Join the sounds together, using action and sounds. Make cards with patterns to represent the different sounds.

Making the Display

- Make a title for the display, based on a favourite character, such as a class soft toy. Place a model house at the top and a model school at the bottom of the display.

- Make a road system of straight roads between the two, using grey or brown paper.

- Put in road features, such as zebra crossings, streetlights and signs.

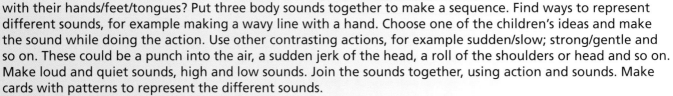

- Add a variety of scenery, such as fields, a farm, a fire station and so on. Make each item interesting by having moving parts, such as doors that open to show what is inside and trees with moving leaves.

- Draw the chosen character at each point, showing whether it likes or dislikes the sound it hears. Decide on a way to show this – it could be a smile or a frown.

- Add numbers at different places. Ask the children to name the sound at each number.

- Draw pictures of things around us that make sounds. Sequence them from quietest to loudest.

- Have a quiet listening time in the classroom. The children close their eyes, put their heads on the table and listen carefully for a minute. List the sounds they hear. Do this at different times of the day and compare the range of sounds. Ask different groups to draw the sounds at different times.

- Make some simple musical instruments, such as shakers and things to bang. Try putting different things into shakers, such as dried peas, rice, breakfast cereal and marbles, and compare the sounds they make. Which makes the loudest/ quietest sound?

- On a chosen musical instrument, find ways to make three different sounds. Make up a sequence of the sounds on the instrument.

- Talk about the importance of listening carefully for safety. Talk about sounds that warn us of danger – fire engines, police cars, ambulances and fire alarms. Link this to a display on traffic sounds.

- Make different sounds with the voice without using words. Make them sound happy, angry, sad or frightened. Which sounds are loud/ quiet, high/low?

- Find out about deafness and how it can stop people from learning to speak well. Learn some of the sign-language alphabet. Add the signs to a familiar song.

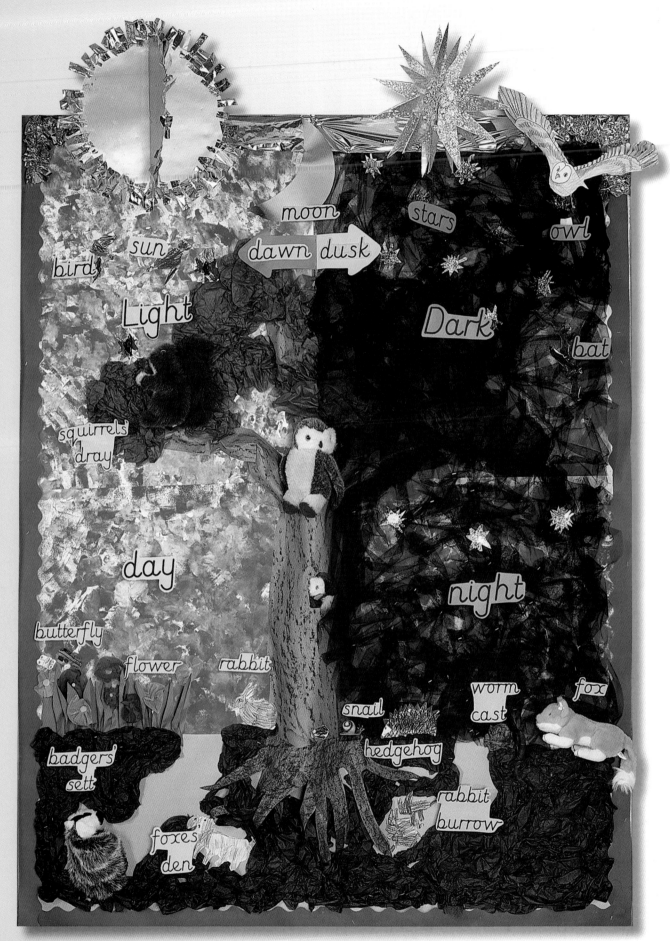

Light and Dark (page 48)